Anoja Kurian

A journey through the DNA barcode

Anoja Kurian

A journey through the DNA barcode

Unlocking nature's secrets: a fundamental guide to DNA barcoding

ScienciaScripts

Imprint

Cover image: www.ingimage.com

This book is a translation from the original published under ISBN 978-3-8416-3673-7.

Publisher:
Sciencia Scripts
is a trademark of
Dodo Books Indian Ocean Ltd. and OmniScriptum S.R.L publishing group

120 High Road, East Finchley, London, N2 9ED, United Kingdom
Str. Armeneasca 28/1, office 1, Chisinau MD-2012, Republic of Moldova, Europe
Managing Directors: Ieva Konstantinova, Victoria Ursu
info@omniscriptum.com

Printed at: see last page
ISBN: 978-620-8-51902-5

Contents

INTRODUCTION

All life forms on Earth exist thanks to a continuous process of evolution. This journey dates back some 4 billion years, beginning with the emergence of single-celled organisms and progressing through the revolution of multicellular life forms. Over time, this evolutionary trajectory has led to the remarkable diversity of life observed today. Charles Darwin's revolutionary theory of revolution, set out in his seminal work "On the Origin of Species" in 1859, revolutionised our understanding of the development of life and profoundly influenced its application in the classification of plant and animal groups.

Systematics, as defined by Mayr in 1969, encompasses the scientific exploration of the diversity of organisms, covering their classification, origins, evolutionary pathways, diversity and geographical distribution. This interdisciplinary field is based on a synthesis of morphological, genetic and evolutionary knowledge of organisms, encompassing both taxonomy and phylogeny. Traditionally, phylogenetics, the branch of systematics responsible for reconstructing the evolutionary relationships between species, relied mainly on morphological characteristics. However, contemporary phylogenetic analyses have expanded to incorporate genetic data, such as DNA or protein sequences, alongside morphological traits. This integrative approach recognises that morphologically similar species are not necessarily closely related genetically. By merging morphological and genetic data, researchers aim to achieve a more complete understanding of complex and enigmatic taxa. Phylogenetic reconstructions depend on precise species identification, which is increasingly supported by molecular data. Similarly, taxonomy, a fundamental component of systematics, involves the meticulous description, naming and classification of organisms. Taxonomic categorisation relies on a wide range of observational data, including morphology, behaviour, genetics and biochemical characteristics, to delimit distinct species.

The concept of biological species defines species as groups of naturally interbreeding populations that are reproductively isolated from other groups.

However, despite its importance, this concept is only one of many proposed definitions of the term 'species'. Various alternative species concepts have emerged over time, each with its own merits and limitations, leading to ongoing confusion and debate about the delimitation of species. Different groups of biologists defend different concepts of species, further complicating the issue and underlining the ambiguity inherent in species delimitation. Even after Mayden's compilation of twenty-four species concepts in 1997, other alternative definitions continue emerge, reflecting the complexity of the issue. The diversity of species concepts often leads to contradictory conclusions about the number species present and their limits. Consequently, the problem of the species concept is intimately linked to the challenge of species delimitation. Organisms may have similarities but belong to different species, or conversely, they may look different but belong to the same species. Furthermore, within a single species, the variability of characteristics can complicate efforts to define species boundaries on the basis of observed data. Consequently, determining species boundaries empirical observations is often a difficult task, requiring careful consideration of multiple factors and often leading to ongoing debate and revision.

Over the last two and a half centuries, taxonomists have formally identified around 1.78 million species of plants, animals and micro-organisms. However, this number is derisory compared with the estimated total number species on Earth, estimated at over 30 million. This vast disparity highlights the urgent need to tackle the complexities inherent in the concept of species, a challenge that requires particular attention.

Relying solely on morphological characteristics for species identification presents considerable obstacles, particularly in cases where organisms exhibit extreme environmental plasticity or homoplasy. Environmental plasticity refers to the ability of organisms to adapt to different environmental conditions, often resulting in variations in their physical characteristics that can confound traditional identification . Homoplasy, on the other hand, refers to the phenomenon in which different species independently develop similar traits, leading to superficial similarities that can obscure true evolutionary relationships. Given these challenges, it is becoming increasingly problematic to accurately delimit species boundaries on the basis of morphological traits alone.

Species nomenclature can become confusing, particularly when the distribution of a species spans several linguistic and cultural groups. General terms often encompass a wide range species, reflecting different linguistic, cultural or developmental nuances. For example, the term 'canes' is commonly used to designate a wide range species. However, this broad categorisation can obscure the distinct identities and ecological roles of individual species within the group. Despite their convenience, general terms can mask important differences in morphology, behaviour or ecological function between species.

In the context of rattan, the use of general terms poses challenges for the accurate identification and classification of species. The diversity of rattan species covered by these terms may present significant variations in their characteristics such as growth habit, leaf morphology or ecological requirements. Therefore, relying solely on general terms may hamper efforts to develop tailored conservation strategies or sustainable management practices for specific rattan species. Furthermore, general terms can obscure the commercial importance of individual species within the rattan group. By grouping distinct species under a single umbrella term, the economic value or unique utility of some species may be overlooked, leading to missed opportunities sustainable use or resource management.

In the field of marine biology, the term 'corals' is often used as a general term for a wide range organisms belonging to the phylum Cnidaria. Although this term is convenient for general discussions, it fails to capture the great diversity and ecological importance of individual coral species. Within the broad category of 'corals', there are a multitude of species with varying morphologies, growth forms and ecological roles. For example, some corals form massive reef structures that provide essential habitat for many marine species, while others exhibit branching or encrusting growth forms. In addition, coral species can differ in their symbiotic relationships with algae, their sensitivity to environmental stressors and their contribution to the functioning of ecosystems.

In the botanical world, the term 'gramineae' is often used as a general term to encompass a wide range plant species belonging to the Poaceae family. Although convenient for general discussion, this term overlooks the diversity of gramineae species with varying growth habits, ecological niches and economic uses. Similarly, in the animal kingdom, the term 'birds' is commonly used as a general term for avian species of various taxa. However, this term fails to capture the rich diversity of bird species, which vary considerably in terms of size, habitat preferences, migratory behaviour and ecological roles. In the field of fungi, the term "fungi" is used as a general term to designate various species of fungi with fruiting bodies. However, within this category, there is a multitude of fungal taxa, each with its own morphology, ecological interactions and medicinal or culinary uses. Furthermore, in the field of marine life, the term 'fish' is often used as a general term for aquatic vertebrates. However, this term encompasses a wide range species covering diverse habitats, behaviours and evolutionary lines, from tiny reef-dwelling gobies to huge tuna roaming the ocean.

Although general terms are a convenient shorthand for discussing broad categories of organisms, they can obscure important distinctions between species, hampering efforts to understand and effectively conserve biodiversity. It is therefore essential to recognise and appreciate the diversity encompassed by these broad terms, and to use more precise language when discussing specific taxa and their ecological roles. Overcoming the use of general terms in taxonomy requires a concerted effort improve identification and classification at the species level. By taking advantage of molecular tools and rigorous taxonomic methods, researchers can unravel the complexities inherent in biodiversity and establish a more nuanced understanding of species boundaries and ecological roles. This approach not only promotes more accurate species identification, but also facilitates targeted conservation efforts and sustainable management practices tailored to the specific needs of each species.

As a result, the need to complement traditional morphological approaches with molecular techniques is increasingly recognised, in order improve species identification and classification. By integrating molecular data with morphological observations, researchers can achieve a more complete understanding of species diversity and evolutionary relationships, thereby advancing our knowledge of the natural world.

The abundance of DNA sequence databases facilitates comparisons between a wide range organisms, providing researchers with invaluable information on evolutionary relationships and facilitating more accurate taxonomic classifications. Molecular data, which lends itself to statistical modelling, provides a robust framework for understanding evolutionary change and precise species identification, thus addressing taxonomic complexities within diverse groups. Given these advantages, numerous molecular approaches have been explored to solve taxonomic challenges, with DNA barcoding emerging as a particularly promising tool in this respect.

CHAPTER 1

DNA BAR CODING

The advent of DNA nucleotide sequence data has led to renewed interest in systematic molecular analysis, offering a promising avenue for accurate identification and classification within various taxonomic groups. In particular, the classification of genera has benefited considerably from these advances. Since Paul Hebert's pioneering proposal of DNA barcoding for the identification of organisms, this technique has gained widespread recognition, particularly in the field of plant species identification. The quest for an ideal DNA barcode persists, motivated by the desire for better discrimination and identification of species using concise DNA segments, characterised by high recovery rates and universality. DNA barcoding is a molecular technique rapid, accurate and automatable species identification using standardised DNA sequences called DNA barcodes, enabling cost-effective species identification. This introductory exploration sets the scene for further exploration of the importance and implications of DNA barcoding in contemporary biological research.

The fundamental principle behind the DNA barcode lies in its ability to uniquely identify an organism at the species level by comparing its individual DNA sequence with a reference library of such sequences. This analogy is similar to the way a supermarket scanner works, using the black stripes of a UPC barcode to identify an item in its inventory by referencing a database. DNA 'barcodes' are used for a variety of purposes, including the identification unknown species or specific parts of an organism, the complete cataloguing of taxa and comparison with traditional taxonomic methods to elucidate species boundaries.

The use of DNA barcodes enables researchers to identify species quickly and accurately by comparing their genetic signatures with those stored in reference databases. This method enables species to be identified even when traditional morphological characteristics are ambiguous or unavailable. Similar to the way a supermarket scanner quickly associates an item with the corresponding database entry, the DNA barcode facilitates the rapid and reliable identification of organisms on the basis of their genetic information.

In addition, DNA barcodes are being used not only to identify known species, but also to discover new ones. By cataloguing as many taxa as possible, researchers can discover previously unknown species and contribute to our understanding of global biodiversity. In addition, DNA barcoding can complement traditional taxonomic approaches by providing molecular evidence to support species delimitation and refine existing taxonomic classifications. Through comparative analyses with traditional taxonomy, DNA barcoding helps to clarify species boundaries and resolve taxonomic uncertainties, thereby improving our knowledge of species diversity and revolution.

In 2003, Paul D.N. Hebert and his colleagues at the University of Guelph, Ontario, Canada, proposed the specific methods and terminology of DNA barcoding in their groundbreaking paper. This standardised approach aimed to identify species and potentially assign unknown sequences to higher taxa such as orders and phyla. Hebert and his team demonstrated the efficiency of the cytochrome c oxidase I (COI) gene, first used by Folmer et al. in 1994. Using published DNA primers, they demonstrated the usefulness of the COI gene as a tool for conducting phylogenetic analyses at the species level, particularly for metazoan invertebrates. The distinct patterns of variation in the COI gene at DNA level make it well suited to distinguishing taxa. Its relative ease of sequence recovery, combined with its variability and conservation between species, offers many advantages. Hebert and his colleagues have coined the term 'barcodes' to describe these profiles and have envisaged the development of a COI database that could serve as the basis for a universal bio-identification system.

The cytochrome c oxidase I (CO1) gene, a mitochondrial DNA (mtDNA) sequence, has been widely recognised as an effective barcode for species identification in various animal groups, as demonstrated by Hebert et al. in 2003. Mitochondrial DNA (mtDNA) has several advantages over nuclear DNA. The mutation rate of DNA is inversely correlated with genome size. As a result, nuclear DNA mutates relatively more slowly than mtDNA. As a result, a much longer nucleotide sequence would be required for nuclear DNA to serve effectively as a barcode capable of distinguishing species, as opposed to the comparatively shorter sequence required with mtDNA. However, in plants, mitochondrial DNA (mtDNA) has low substitution rates and undergoes rapid changes in gene content and structure. As a result, this region appears unsuitable for DNA barcoding. As a result, plant barcoding typically uses chloroplast DNA regions, such as rbcL, matK and trnH-psbA, due to their relatively high rates of sequence variation and ease amplification. The development of standardised protocols and databases, such as the Barcode of Life Data Systems (BOLD), has facilitated the application of plant barcoding in various fields, including biodiversity assessment, conservation and medicinal plant authentication.

Numerous international organisations, including the Consortium for the Barcode of Life (CBOL) and the Barcode of Life Data Systems (BOLD), have been established to advocate and promote DNA barcoding as a universal standard for species identification. These organisations play a key role in advancing research, facilitating collaboration and promoting the widespread adoption of DNA barcoding across diverse scientific disciplines and geographic regions.

The Consortium for the Barcode of Life (CBOL), for example, is an international initiative dedicated to advancing the DNA barcode as a universal standard for species identification. Created in May 2004 with the support of the Alfred P. Sloan Foundation, CBOL has over 130 member organisations in more than 40 countries. The CBOL secretariat office is based at the National Museum of Natural History, Smithsonian Institution, in Washington, DC.

CBOL actively promotes DNA barcoding through various channels, including workshops, working groups, international conferences and awareness-raising meetings in developing countries. These initiatives aim to raise awareness of barcoding and foster collaboration between scientists and stakeholders around the world. In addition, CBOL facilitates planning meetings for barcode projects and produces outreach materials to disseminate information about the benefits and applications of DNA barcoding. An important step taken by CBOL is the development a data standard for DNA barcoding, which has been endorsed by major repositories such as GenBank, the European Bioinformatics Institute and the Japanese DNA Database. This standardisation improves interoperability and accessibility of DNA barcode data, facilitating its integration into wider scientific research and conservation efforts.

CBOL's advocacy efforts extend beyond the scientific community to involve government agencies and international organisations involved in agriculture, environmental conservation and biodiversity management. In advocating the adoption of DNA barcodes in these sectors, CBOL seeks to demonstrate the practical utility of barcodes in a variety of areas, including species monitoring, trade regulation (such as through CITES) and sustainable agricultural practices. Thanks to its multi-dimensional approach, CBOL has played a central role in promoting the widespread acceptance and implementation of DNA barcoding as a powerful tool for species identification and conservation.

Similarly, BOLD operates as an online workshop that supports the collection, management, analysis and application of DNA barcodes. Housed at the University of Guelph, BOLD provides researchers with a comprehensive platform for downloading, storing and analysing DNA barcode data, facilitating the integration of DNA barcoding into various research projects and applications. With its user-friendly interface and extensive database of DNA barcode sequences, BOLD researchers with access to valuable resources and tools for species identification, taxonomic research and biodiversity monitoring. The Barcode of Life Data Systems (BOLD) serves as an Internet platform facilitating the collection, organisation, analysis and application of DNA barcodes. Led by Hebert, who is director of the Biodiversity Institute of Ontario and the Canadian DNA Barcoding Centre, as well as the International Barcode of Life Project (iBOL), all based at the University of Guelph, BOLD is also located here. institution. Together, organisations like CBOL and BOLD are playing a key role in advancing the field of DNA barcoding, promoting interdisciplinary collaboration and improving our understanding of global biodiversity.

CHAPTER 2

DNA BARCODES

DNA barcoding involves the use of distinctive nucleotide sequence patterns derived from small fragments of DNA, typically between 400 and 800 base pairs, to serve as specific reference collections sample identification called DNA barcodes. DNA barcodes are therefore short, standardised DNA sequences used as unique identifiers for species. These sequences, usually derived from specific regions of an organism's genome, serve molecular labels that can be used to distinguish between different species. DNA barcodes are used in a variety of fields, including biodiversity assessment, species identification and conservation biology, providing a rapid and accurate method for classifying and cataloguing organisms based on their genetic information. Initially, the main aim of the DNA barcode was establish comprehensive online libraries containing sequences of all known species. These libraries would then function as standardised references against which the DNA barcodes of any identified or unidentified specimen could be compared and matched.

The essential characteristics of an ideal DNA barcode include universality, specificity, variation and ease of use. Universality ensures that the selected DNA fragment can be amplified and sequenced in a wide range species, making it easy to apply in a variety of taxa. Specificity refers to the ability of the barcode to accurately distinguish between closely related species or individuals within a species. Variation is essential for capturing genetic diversity within and between species, enabling meaningful differentiation. Finally, ease of use ensures that DNA barcoding can be efficiently and reliably generated using standard laboratory techniques, enabling widespread adoption and application in a variety of scientific and practical contexts. The barcode must contain a limited number insertions or deletions facilitate sequence alignment and ensure uniformity. In addition, its mutation rate must be sufficient to create a "barcode gap", in which the highest intraspecific variation is less than the lowest interspecific variation. The "barcode gap" refers to the difference between the maximum genetic variation found in individuals of the same species and the minimum genetic variation observed between individuals of different species.

Barcodes in animals

Mitochondrial cytochrome c oxidase subunit I (COI) is the predominant DNA barcode marker for animal identification due to its widespread use and high conservation in species dependent on oxidative phosphorylation for metabolism. Numerous studies have highlighted effectiveness of COI-based DNA barcoding for delimiting various animal species, highlighting the significant rates of sequence variation at the species level and the constraints of intraspecific divergence within COI sequences. For the analysis of fresh, well-preserved animal tissues, experts recommend the use of a complete barcode, such a 658 bp region of the COI gene, as this allows feasible PCR amplification and sequencing. Folmer's development of 'universal' primers to amplify the COI gene in metazoan invertebrates revolutionised the DNA barcoding process, enabling rapid and simple barcoding and phylogenetic analysis. However, the proliferation of barcoding studies has led to a higher frequency of incorrectly published sequences. This is often due to the unintentional amplification of non-target taxa and insufficient analysis of the sequences obtained. As a result, some sequences deposited in genetic databases are wrongly labelled as coming invertebrates they are in fact bacterial sequences. To overcome the constraints associated with Folmer primers, researchers have developed alternative primer sets. Since 2013, the Barcode of Life database (BOLDSystems) has catalogued around 418 distinct primers designed to target COI sequences from various taxa. In addition, some scientists have experimented with significantly reducing primer annealing temperatures to mitigate amplification problems.

In addition to COI, other genetic markers such as 16S rRNA and ITS regions have been used as DNA barcodes for certain groups animals. These markers are particularly useful in cases where COI may not provide sufficient resolution, such as invertebrates or closely related species.

However, animal tissues often undergo processing before being used as traditional medicines (TMs), and certain processing methods such as sun-drying, frying or boiling can induce DNA degradation, complicating PCR amplification of complete barcodes. As the feasibility of amplification improves with shorter sequence lengths, there is growing interest in using a 'minibarcoding' approach, which uses abbreviated regions in the standard barcode. Recent research indicates that a range of minibarcodes can be efficiently amplified and sequenced from a variety of processed products, including TMs and foods, providing adequate sequence information essential for species identification. In some research projects, DNA barcoding has provided a valuable tool for the discovery of potential cryptic species, synonymous taxa or species at risk of extinction.

In addition, it facilitated the matching of adult specimens with their immature counterparts. However, the accuracy of DNA barcoding in resolving higher taxonomic groups was not as precise as its effectiveness at the species level. Nevertheless, DNA barcoding shows promise as a method species identification and could become a widely adopted standard tool in taxonomic and biodiversity research.

Barcodes in plants

Plant barcoding involves the use of DNA sequences from specific regions of the plant genome to identify and classify plant species. This approach provides a rapid and accurate method for distinguishing between different plant taxa, overcoming the challenges associated with traditional morphology-based identification, particularly in the case cryptic species or morphologically similar taxa.

In plants, mitochondrial DNA (mtDNA) has low substitution rates. As a result, this region is considered inadequate for DNA barcoding applications. Recognising the limitations of mtDNA in plants, the Consortium for Barcoding of Living Plants (CBOL) Working Group proposed an alternative approach. They recommended the use of seven chloroplast genomic regions for species identification in the plant kingdom. These regions comprise four coding gene segments (matK, rbcL, rpoB and rpoCl) and three non-coding spacers (atpF - atpH, psbA-trnH and psbK - psbI). By focusing on chloroplast DNA, which has higher substitution rates and greater stability of gene content and structure than mitochondrial DNA, the CBOL Plant working group sought to overcome the challenges associated with barcoding plant DNA.

Chloroplastic regions

The coding and non-coding regions of plastids, such as the intergenic spacers rbcL, matK and psbA-trnH, are commonly used for plant barcoding. The rbcL gene encodes the large subunit of rubilose-1,5-bisphosphate carboxylase/oxygenase (RUBISCO), an enzyme essential for photosynthesis. Although suggested by CBOL as a potential barcode for plants, rbcL alone is not particularly effective at distinguishing species, leading many research groups to recommend its use in combination with other markers.

The matK gene encodes maturase kinase, which plays a role in the splicing of type II introns from RNA transcripts. Because of its rapid evolution, some groups recommend using matK either as a single barcode or in combination with other markers. However, PCR amplification failure has been observed in some species, showing a limitation of the matK region. Because of their high species discrimination, universality and ease of PCR amplification, the combination of rbcL and matK has been approved by CBOL as a standard two-locus barcode for plants.

Similarly, other chloroplast regions such as rpoB, rpoCl, rpoC2, accD, ycf5 and ndhJ have been suggested by various working groups for plant barcoding. The rpoB, rpoCl and rpoC2 genes code for RNA polymerase subunits and are commonly used in phylogenetic studies due to their high substitution rates. Although these genes are highly universal and produce high quality sequences, they have been rejected as unique barcodes by the CBOL Plant Working Group because of their poor species discrimination.

The accD gene, which codes for the ʙ-carboxyl transferase subunit of acetyl- CoA carboxylase, is found in all flowering plants but is completely absent in gramineae. It evolves five times faster than rbcL in plant taxa and has been proposed as a barcode. The Royal Botanic Gardens, Kew, have proposed the short, single-copy ycf5 gene, which codes for a protein containing 313 amino acids, as a barcode region. Although ycf5 is conserved in all land plants, it is rarely used due to low universality and difficulties sequence alignment.

The ndhJ gene codes for a subunit of NADH dehydrogenase, involved in respiratory processes and part of a gene complex made up of 11 genes (ndhA - ndhK). Although ndhJ has been tested for barcoding studies, it is absent in some gymnosperms such as Pinus, Cuscuta and Gnetales, but present in other gymnosperms. addition, amplification difficulties have been reported for ndhJ in orchids. Although few studies have recommended ndhJ as an additional barcode because of its great ability to discriminate species in hepatics, pteridophytes and monocotyledons, CBOL has not accepted it, probably because of its absence in certain groups such as conifers. In addition, some studies have found that ndhJ has little discriminatory power.

Intergenic spacers such as psbA-trnH, atpF-atpH and psbK-psbI are also used as plant barcodes. Due to its high discriminatory power between species, the psbA-trnH spacer is widely used in plant DNA barcoding. The genome size of psbA-trnH varies from around 300 bp to 1,000 bp, which can sometimes lead to bidirectional sequencing problems. This region is often combined with rbcL and matK to improve species discrimination.

The intergenic spacer atpF-atpH encodes the ATP synthase subunits CFO I and CFO III, respectively. Although this spacer amplifies readily, sequence alignment can be difficult due to length variations, leading CBOL to recommend it only as an additional locus. The psbK and psbl genes encode two low molecular weight polypeptides, K and I, respectively, of photosystem II. Unlike atpF-atpH, psbK-psbl exhibits simple sequence alignment and PCR amplification. However, due to the difficulties encountered in obtaining unambiguous bidirectional sequences, it is also considered as an additional barcode by CBOL, despite its high species discrimination power.

In addition, the trnL (UAA) - trnF (GAA) locus, although not recommended by CBOL, has been suggested as a barcode by some researchers. This locus is advantageous because of its secondary structure, which comprises alternating conserved and variable regions, and its utility in highly degraded tissues. This feature makes trnL (UAA) - trnF (GAA) particularly useful in barcoding studies involving degraded samples.

Due to the limited variation in plastid DNA, several regions are often selected as barcodes using a multi-level approach. The use of rpoC1, rpoB and matK, as well as the combination of rpoC1, matK and psbA-trnH, as efficient markers for land plant barcoding has been proposed. Instead of simply adding more locations to increase the number of regions, a multi-level strategy was suggested. This using a first level coding region common to all land plants and a second level of non-coding regions to provide information at the species level. Multi-gene approaches have become common practice in many plant species.

Nuclear region

The nuclear ribosomal DNA (rDNA) cistron is a multigenic family responsible for coding the nucleic acid core of ribosomes. This family evolves in a concerted manner, resulting in high degree of sequence homogeneity among rDNA copies within a species, while still exhibiting significant differences between species. This characteristic gives rDNA the qualities of an ideal DNA barcode.

Internal transcribed spacer (ITS) regions of rDNA demonstrate superior discriminatory power to plastid regions, particularly at lower taxonomic levels. ITS1 and ITS2 regions can be amplified separately using primers anchored in conserved regions. This facilitates amplification of these regions even degraded DNA , making ITS particularly useful for species identification.

Despite their usefulness, ITS regions can show intra-individual variation. These variations can occur as a result of recent hybridization events, line sorting, recombination between copies, high mutation rates and the formation of pseudogenes in cistrons. Nevertheless, the nuclear ribosomal ITS region (nrITS) is still considered to be a serious candidate for plant barcoding due to its multiple copies in each cell, the availability universal primers, its high universality and its good species discrimination power.

The application of nrlTS to plant identification has been particularly successful in identifying medicinal plants and their close relatives. This success highlights the usefulness and reliability of nrlTS for distinguishing closely related plant species, making it a valuable tool in the field of plant barcoding.

Low copy nuclear regions

The slow evolution of chloroplast DNA (cpDNA) and the incomplete concerted evolution of nuclear ribosomal DNA (nrDNA) have highlighted the need for additional markers for species identification. Studies have shown that plastid data often fails to accurately define species boundaries in groups such as willows, wild potatoes, sedges and bryophytes. This problem is often due to factors such as widespread hybridisation, introgression or incomplete sorting of lines.

Since low copy number nuclear genes evolve much more rapidly than the plastid genome and are common in rapidly diversifying lines, these genes are now being considered as potential barcodes for species discrimination. Research has shown that low copy number nuclear regions can serve as efficient DNA barcodes, as demonstrated by plant groups such as *Clermontia* (Campanulaceae) and *Cyrtandra* (Gesneriaceae), where plastid regions exhibit a slow rate evolution. This approach offers a promising alternative for distinguishing species in cases where traditional plastid markers are insufficient. Although it is a low copy number nuclear gene region, RPB2 has been shown to be an effective barcode for distinguishing Calamus species. RPB2, a low copy number nuclear gene region, has gained importance in the field of plant DNA barcoding due to its effectiveness in species delimitation. Unlike plastid regions, which often have slow rates of evolution, RPB2 evolves more rapidly, making it a valuable tool for distinguishing closely related species. This genetic region has demonstrated its usefulness in accurately identifying species of the Calamus genus and potential application in other genera of the Palm family. The increased resolution provided by RPB2 helps to overcome the challenges associated with hybridisation, introgression and incomplete line sorting, which can obscure species boundaries when relying solely on plastid or nrDNA markers. Thus, RPB2 represents a promising addition to the suite of genetic markers used for accurate and reliable species identification in slowly evolving plants. It also shows promise as a potential barcode for other Palm genera.

CHAPTER 3

ADVANCES IN DNA BAR CODING

The future of DNA barcoding offers exciting opportunities for further progress. With the continued development of sequencing technologies, such as next-generation sequencing and nanopore sequencing, the cost and time required for DNA sequencing is likely to continue to decrease. This will make it possible to generate large quantities of sequence data from various organisms, facilitating the complete identification of species and the exploration of biodiversity on a global scale.

Although universal DNA barcoding is not fully effective in identifying species across the entire tree of life, researchers have adopted DNA barcoding as a valuable tool in their fields, leading to the introduction of many new barcoding regions. A significant advance is the use of complete plastid genome sequences as DNA barcodes. The decreasing costs and increasing capabilities of new generation sequencing technologies have made whole plastid sequencing more feasible. As plastid haplotypes often do not perfectly match species boundaries, sequencing the entire plastid genome can provide clearer and more definitive barcoding results. The development of new-generation sequencing technologies will eventually make it possible to acquire large quantities of sequence data from many individuals, potentially resolving the problems associated with single-copy genes. This global approach could make it possible to delimit species boundaries as thoroughly as possible using genetic data.

In addition, the emergence of mini-barcodes, which are shorter DNA sequences, offers a more cost-effective and efficient approach to DNA barcoding, particularly for large-scale species . Traditional DNA barcoding typically uses a 650 bp DNA segment of the mitochondrial COI gene for species identification in animal groups, while fragments of similar size from chloroplast genes are suggested as barcode markers for plants. Although PCR amplification and sequencing of this 650 bp fragment is generally reliable in freshly collected and well-preserved samples, obtaining a complete barcode may prove difficult in older museum specimens or those preserved in DNA-unfriendly substances such as formalin. This limitation also applies to processed biological materials such as food products, pharmaceuticals and nutraceuticals, hindering effective DNA-based authentication and testing. To solve this problem, shorter DNA sequences, known as mini-barcodes, have been successfully recovered and have demonstrated their ability to effectively identify the majority of specimens at the species level. In addition, these short DNA regions can be processed using high throughput sequencing platforms, offering a cost-effective and comprehensive approach to large-scale species . Standardised universal primers make mini-barcodes highly accessible, enabling researchers to rapidly analyse DNA and identify species with greater accuracy. The combination of mini-barcode properties and the availability standardised universal primers make them a viable option for DNA barcoding analysis of museum samples, as well as for applied diagnostic and environmental biodiversity analysis.

With advances in high-speed sequencing technologies, obtaining complete chloroplast genomes has become more accessible on a large scale and at reduced cost. This has stimulated studies on systematic analyses using chloroplast genomes in plant genera such as *Epimede, Paris,* and *Sanguisorba*. Due to the limited discriminatory power of conventional molecular markers in plants and their closely related species , researchers suggested using the entire chloroplast genome as a super-barcode to distinguish closely related species. Compared with the commonly used genus-specific DNA barcodes, the chloroplast genome exhibits greater variation, leading to significantly improved resolution in phylogenetic analyses. As a result, the chloroplast genome has found many applications in phylogenetic studies, plant population analysis and plant identification. Phylogenetic trees constructed on the basis of complete chloroplast genomes tend to have higher support rates and discrimination power. Consequently, some researchers have proposed using the entire chloroplast genome as a super-barcode to accurately identify closely related plant species.

As super-barcoding gains ground, a major challenge is to balance the high cost with the amount of cp genome sequences available in databases, particularly when replicate samples are scarce. Therefore, the use of super barcodes is not always recommended, particularly when commonly used DNA barcodes can adequately fulfil identification purpose. As a result, specific barcodes are often developed and used for plant identification when standard barcodes fail to provide accurate authentication, and superbarcodes are deemed excessive for the needs of certain experiments, particularly within specific taxonomic groups.

Nevertheless, the advantages of the super-barcode are clear in situations where traditional DNA barcodes struggle to accurately identify species at lower taxonomic levels.

Although early studies demonstrated the potential of the super-barcode, there is evidence to suggest that the super-barcode does not always allow the identification of all plant species. As a result, some researchers have refocused on nuclear genes. A combination of plastid genome and rDNA has gradually been accepted for species identification, with the genome template serving as a universal extended barcode. However, earlier analyses were carried out at genomic level. Whole cp genomes and 6 k bases of nuclear ribosomal DNA were used as an 'ultra-barcode' to differentiate cocoa species, including several individuals of T. cacao and one individual of T. grandiflorum. Their results showed that ultra-barcoding is a feasible approach for accurately distinguishing varieties and even individuals of different genotypes within cocoa. Genome skimming, based on low-coverage sequencing of total DNA, can recover complete sequences of the plastid genome, eliminating the need obtain whole cp genome sequences, as required by traditional super-barcoding. Genome skimming was performed to obtain nearly complete plastid genomes and rDNA for rhododendron species using a large number samples. They compared the success of discrimination with that of multi-locus DNA barcoding and attributed the significant increase in discriminatory power to the large amount of data on the plastid genome.

Pangenome-based barcoding represents an innovative approach to DNA barcoding that exploits the concept of the pangenome, which encompasses the entire genomic repertoire of a species, including core genes shared by all individuals and accessory genes present only in a subset of individuals. Unlike traditional barcoding methods that focus on a single gene or a small set of genes, pangenome-based barcoding takes into account the entire genomic diversity within a species. One of the main advantages of pangenome-based barcoding is its ability to capture more fully the genetic variability present within species. By incorporating information from both core and accessory genes, pangenome-based barcodes can provide a more robust and accurate identification of species, particularly in cases where traditional barcodes fail to differentiate between closely related taxa.

In addition, pangenome-based barcoding has the potential to uncover new genetic markers that may be overlooked by traditional barcoding methods. By exploring the entire genomic landscape a species, researchers can identify unique genomic regions that exhibit high levels of variability and can serve as informative barcodes for species identification.

In addition, the concept of super barcodes, based on whole plastid sequences, the barcodes of the pangenome, is showing promise for discrimination between closely related species. As sequencing technologies progress, whole genome sequencing and super barcoding could become increasingly feasible and could become the preferred choice for plant identification.

Overall, future advances in DNA barcoding should revolutionise our understanding of biodiversity, improve species identification efforts and support various fields such as conservation biology, ecology and forensic medicine.

CHAPTER 4

DNA BARCODING METHODOLOGY

The DNA barcoding methodology is straightforward, the process being broadly similar for plants and animals, with the exception of differences in DNA isolation techniques and the specific primers used.

Once a suitable DNA segment has been selected for analysis, it must first be extracted from the sample and amplified by polymerase chain reaction (PCR) specific barcode primers. The amplified segment of the gene is sequenced and this sequence, the 'barcode', is then compared to existing barcodes or to material from reference specimens after deposit in the NCBI. The genetic distance correction of the Kimura2 parameter (K2P) is used to quantify sequence divergences between individuals, as it is the most effective model when distances are small.

DNA extraction

Obtaining DNA with high yield and quality is essential for successful DNA barcoding. This requires careful sampling and thorough homogenisation, particularly for animal tissues with limited DNA content. The DNA extraction process typically involves several key steps, including tissue lysis, removal of impurities and DNA precipitation. Various classical methods exist for extracting DNA from animal tissues, such as sodium dodecyl sulphate (SDS) extraction, guanidinium thiocyanate-phenol-chloroform extraction, silica matrix purification and magnetic bead purification. Many laboratories have adapted these traditional methods to improve efficiency of DNA extraction. For example, a cost-effective and user-friendly protocol has been developed using SDS and proteinase K for tissue lysis, followed by silica-based purification using glass fibre filter plates. In addition, several of these protocols have been commercialised in kit form, offering standardised DNA solutions for different research needs. Given the distinct characteristics of different animal tissues, it is crucial to select appropriate extraction methods or commercial kits tailored to the specific sample type.

In the case of plants, the CTAB method is generally used and is described in detail below.

The isolation of plant DNA using the CTAB (cetyltrimethylammonium bromide) method involves several key steps to efficiently extract DNA from plant tissues. Firstly, sample preparation is crucial. Fresh plant tissue samples are collected, ensuring that they are young and healthy for optimal DNA extraction. It is essential remove any dirt, debris or contaminants from the surface of the plant tissue before proceeding with the extraction process. Next, the plant tissue is ground or homogenised to release the cellular contents, including the DNA. This can be achieved by grinding the tissue in a mortar and pestle with liquid nitrogen or by using a homogeniser to create a fine powder or homogenate.

The CTAB extraction buffer is then prepared. This buffer usually contains CTAB, a detergent that helps to break down cell membranes and release DNA, as well as other components such EDTA and Tris-HCl to prevent DNA degradation and adjust the pH. Once the buffer has been prepared, crushed or homogenised plant tissue is added to the CTAB extraction buffer and the mixture is incubated at a high temperature (usually around 60-65°C). This step facilitates cell lysis, allowing cellular components, including DNA, to be released into the solution.

After cell lysis, a chloroform/isoamyl alcohol solution is added to the lysate to separate the aqueous phase containing the DNA from the proteins and other cell debris. The mixture is then centrifuged to separate the phases, with the DNA remaining in the upper aqueous layer. The aqueous phase containing the DNA is transferred to a new tube and cold isopropanol or ethanol is added to precipitate the DNA. After centrifugation, the DNA pellet is washed with cold ethanol to remove residual contaminants, then air-dried or briefly centrifuged. Finally, DNA pellet is suspended in a suitable buffer or distilled water for downstream applications. The quality and quantity of the extracted DNA is assessed by spectrophotometry or gel electrophoresis to ensure its suitability for subsequent analyses, such as PCR amplification for DNA barcoding. Overall, the CTAB method is widely used for the isolation of plant DNA due to its efficiency in high quality DNA suitable for a variety of molecular biology applications. Several kits are also now available on the market for DNA isolation.

Selecting the barcode region

The selection of an appropriate DNA barcode region is essential for the accurate identification and classification of plant and animal species. In animals, the mitochondrial region of cytochrome c oxidase I (COI) is the most commonly used barcode because of its high rate of sequence variation, which facilitates species discrimination. Extensive studies have demonstrated the capacity of the COI region for animal taxonomy. However, the COI barcode has several limitations. For example, it has been shown to offer insufficient and unreliable discrimination for certain species in the classes Gastropoda and Anthozoa. This has led researchers to explore other genetic markers such as mitochondrial cytochrome b (Cytb), 16S rRNA, 12S rRNA and the internal ribosomal transcribed spacer (ITS) for animal barcoding.

For example, the ITS2 region can achieve a high identification rate of 91.7% at species level among 12,221 animal types recorded in GenBank and can differentiate species such as Argasidae, which are not reliably identified by the COI barcode. Mini-barcodes, shorter regions of DNA, are also used, particularly for highly processed samples where DNA degradation is a problem. Although barcodes generally have a higher amplification success rate than full barcodes, their efficiency is positively correlated to the length of the barcode. As a result, lengths are generally kept above 100 bp. For example, a 250 bp region of 16S rRNA has been successfully amplified from various medicinal preparations and food products, enabling correct identification of animal species.

In plants, barcode regions can be selected from chloroplast, nucleus or low copy number genes. The choice of barcode region depends on the specific plant species under study. Chloroplastic regions such as rbcL, matK and the psbA- trnH intergenic spacer are commonly used due to their universality and relatively high discriminatory power. However, nuclear regions, including ITS regions, are also important for plant barcoding, particularly for closely related species for which chloroplast regions may not provide sufficient resolution.

Single DNA barcodes can sometimes provide insufficient or false identification hybrid species or species with high genetic diversity. To solve this problem, multi-locus barcodes, which involve use of several genetic regions, can improve the accuracy and sensitivity of identification. The Traditional Chinese Medicine (TCM) DNA Barcode Guidelines for Molecular Identification, included in the Appendix of the Chinese Pharmacopeia, recommend complete animal identification using COI and ITS2 barcodes.

The selection of appropriate barcode regions is therefore essential for accurate and reliable species identification. Although COI remains the dominant barcode for animals, its limitations necessitate the use additional markers or multilocus approaches. In plants, the choice of barcode region must be adapted to the species and the specific requirements of the study, thus guaranteeing the most accurate taxonomic resolution possible.

PCR amplification of the Barcode region

The polymerase chain reaction (PCR) is a fundamental molecular biology technique used to amplify specific DNA sequences. PCR is a powerful and versatile technique that has revolutionised molecular biology, enabling the detection, cloning and analysis of specific DNA sequences from minute quantities of raw material.

The PCR method involves several key steps, each carried out in a thermocycler that precisely controls the temperature at each stage of the reaction. Here, this is used to amplify the specific primers used for DNA barcoding. In the case of animals, the CO1 gene primer is used. In plants, chloroplast, nuclear or low copy number primers are used. The universal primers commonly used for plants encode matK, rbcL, rpoB and rpoCl, and three non-coding spacers (atpF - atpH, psbA-trnH and psbK - psbl. For fungi, ITS regions are recommended for better species resolution. The PCR method involves preparation of the reaction mixture.

Preparation of the reaction mixture :

The first step in PCR is to prepare the reaction mixture. This mixture comprises the template DNA, which contains the target sequence to be amplified; two primers (forward and reverse primers), short single-stranded DNA sequences complementary to the regions flanking the target sequence; deoxynucleoside triphosphates (dNTPs), the building blocks for the synthesis of new DNA strands; a DNA polymerase enzyme, usually Taq polymerase, which is thermostable and can withstand the high temperatures used in PCR; and a buffer solution that provides the ions and pH required for the reaction. Magnesium ions are also included, as they are essential cofactors for the DNA polymerase enzyme. The PCR steps include the following:

1. Denaturation :

The PCR reaction begins with the denaturation step. The reaction mixture is heated to around 94-98°C for 20-30 seconds. This high temperature causes the double-stranded DNA to separate into single strands by breaking the hydrogen bonds between the complementary bases. This step is crucial because it allows the primers to access the single-stranded DNA templates.

2. Annealing :

After denaturation, the reaction mixture is cooled to between 50 and 65°C for 20 to 40 seconds. This temperature is specific to the primers used and is generally a few degrees below their melting temperature (Tm). During hybridization step, the primers bind or hybridize to their complementary sequences on the single-stranded DNA templates. Appropriate primer design is essential to ensure specificity and efficient binding to target sequences.

3. Prolongation :

The final stage of the PCR cycle extension or elongation, where the temperature is raised to around 72°C, which is the optimum temperature for Taq polymerase activity. During this step, the DNA polymerase enzyme synthetises new strands of DNA by adding dNTPs to the 3' ends of the primers, thus extending the DNA sequence in the direction of the target sequence. The length of this step depends on the length of the DNA fragment being amplified, typically 1 minute per kilobase of DNA.

Cycling:

The denaturation, annealing and extension steps are repeated for 25 to 35 cycles.

Each cycle doubles the amount target DNA, leading to exponential amplification of the specific DNA sequence. The number of cycles required depends on the initial amount template DNA and the desired amount of amplified product. After the last cycle, final extension step is usually performed by incubating the reaction mixture at 72°C for 5-10 minutes. This ensures that any remaining single-stranded DNA is fully extended and that all new DNA strands are completely synthesised. Once the PCR is complete, the reaction mixture is cooled to 4°C or room temperature. The amplified DNA, also known as the PCR product or amplicon, can then be analysed using various techniques such as gel electrophoresis, which allows the DNA fragments to be visualised to confirm the success of the amplification.

Sequengage of the region amplified

Sanger didesoxy sequencing, a traditional and widely used DNA sequencing method, is capable of generating sequencing reads of up to 1,000 base pairs in length. This method is particularly useful for small-scale single-species DNA barcoding due to its reliability and accuracy. However, the low throughput of Sanger sequencing limits its effectiveness for large-scale projects, where the simultaneous analysis of many samples is required. This makes it less suitable for complex or high-volume sample types in which several species may be present.

In contrast, next-generation sequencing (NGS) technology is revolutionising DNA sequencing by enabling the parallel sequencing of multiple DNA fragments in a single reaction. This high-speed capability significantly speeds up the process and enables large datasets to be analysed, making it ideal for extensive barcoding projects. One of the first commercially available NGS platforms, the pyrosequengage 454, has demonstrated the utility of this technology by efficiently analysing a variety of complex mixtures, including environmental samples, food products and medicinal preparations. The ability to efficiently process mixed samples makes NGS a powerful tool in biodiversity studies and quality control in a variety of industries.

Recent advances have led to the development of several benchtop sequencers designed for routine use in the laboratory. Systems such as the Roche 454 GS Junior, Ion Proton, Illumina MiSeq and MiniSeq have made high throughput sequencing more accessible and practical for everyday applications . These platforms are particularly useful for projects requiring fast turnaround times and high data accuracy, extending the scope of DNA barcoding and other genomic analyses.

Despite the many advantages of NGS, it is not without its challenges. NGS platforms can produce sequencing errors, requiring rigorous quality control measures. Prior to barcode analysis, it is crucial to perform quality filtering and slicing of the raw reads to remove any erroneous data. This step ensures the reliability of the sequencing results, which is essential for accurate species identification and other downstream analyses. As NGS technology continues to evolve, these quality control measures are likely to become more refined, improving the accuracy and usefulness of DNA barcoding in a variety of scientific and industrial fields.

Sequence analysis

Raw sequences were manually edited using BioEdit and aligned using the default alignment parameters in CLUSTAL X. Refined refined were were confirmed via BLASTn (//blast.ncbi.nlm.nih.gov/Blast.cgi) against the NCBI GenBank nucleotide database. The Basic Local Alignment Search Tool (BLAST) is a similarity-based algorithm, which matches the query sequence to those in the reference databases, then provides a similarity score based on the part of the query aligned to the reference.

Accurate species identification relies heavily on the availability of reference sequence data stored in various public databases. The main repositories include GenBank, BOLD, the Medical Device DNA Barcode Database (MMDBD), the International Nucleotide Sequence Database Collaboration (INSDC) and the Barcode Indexing System (BIN). These databases provide a multitude of sequences attributed to specific taxa, facilitating the comparative analysis of genetic variation.

GenBank is a database widely used in barcoding studies, hosting a vast collection of over one terabase of sequence data with broad taxonomic coverage. BOLD has collected more than two million COI sequences around 170,000 species, demonstrating its global reach. INSDC also offers a substantial record of Cytb and ITS sequence data.

MMDBD specialises in barcode information on medicinal plants and animals, covering more than 1,700 species listed in various pharmacopeias, including Chinese Pharmacopeia and American Herbal Pharmacopeia. Notably, MMDBD also provides sequence information for common adulterants and substitutes, which is crucial for guaranteeing the authenticity of medicinal materials.

Data analysis

Three separate statistical analyses (distance-based, similarity-based and tree-based methods) are used to assess species discrimination within a genus.

In the distance-based method, intra- and interspecific pairwise distances are calculated using the two-parameter Kimura model (K2P). This approach identifies the most efficient region for barcoding by analysing the barcoding gap, which represents the difference between intra- and inter-species variability.

The similarity-based method compares each sequence with all other sequences in the aligned dataset to identify the closest match, using criteria such as 'best match' and 'best close match. This method corroborates the results of the distance-based analysis of the optimal region of the barcode.

The tree construction method generates a phylogenetic tree to graphically represent the relationships between sequences. This visualisation plays a key role in assessing the ability of specific markers to distinguish closely related species. The construction of a phylogenetic tree makes it possible to identify unknown samples and resolve taxonomic ambiguities, thus providing a clear overview of the genetic relationships within the genus. Several hierarchical clustering algorithms are used to construct phylogenetic trees, including neighbour-joining (NJ), maximum likelihood (ML), maximum parsimony (MP) and Bayesian inference (BI). Using a combination of these algorithms can result in a more reliable identification than using a single method. Various commercial tools are available for constructing and visualising these trees, such as MEGA, PHYLIP and PAUP. These tools facilitate the implementation of different clustering algorithms and help to visualise the resulting phylogenetic relationships, thereby improving the accuracy and reliability of species identification.

CHAPTER 5

DNA BAR CODING APPLICATIONS

DNA barcodes have a wide range applications in a variety of fields. One crucial area where DNA barcoding is playing an important role is in the conservation of natural resources. By accurately identifying species through their DNA, conservation efforts can be better targeted to protect and conserve biodiversity, particularly for endangered and threatened species. In addition, DNA barcodes contribute to the protection of threatened species by enabling the identification of illegally traded or harvested specimens, thus helping law enforcement agencies to combat wildlife trafficking and poaching.

In agriculture, DNA barcodes play a key role in pest control by rapid and accurate identification of insect pests and pathogens. This facilitates the implementation of targeted pest management strategies, minimising crop damage and reducing the need for broad-spectrum pesticides.

In addition, DNA barcodes are invaluable for identifying disease vectors, such as mosquitoes and ticks, thereby contributing to the surveillance and control of vector-borne diseases. By identifying the species responsible for transmitting pathogens, public health authorities can implement targeted control measures to reduce disease transmission.

In environmental monitoring, DNA barcoding is a powerful tool for assessing water quality and ecosystem health. By analysing the diversity and composition of aquatic organisms, scientists can assess the impact of pollution and other environmental stressors on aquatic ecosystems. However, with the DNA barcoding method, species identification can be carried out without any prior knowledge of the samples being studied. This approach is particularly useful for recovering DNA barcodes mixed environmental samples or for identifying problematic specimens whose morphology is unclear, especially in cases where the material is damaged or only partial body parts are available. Therefore, in such scenarios, the use universal primers, such as Folmer primers, remains a viable solution. However, it is important to note that these primers can also co-amplify nuclear mitochondrial pseudogenes (numts) or 'COI-like' sequences, as well as microorganisms. This can potentially lead to an overestimation of species diversity and have an impact on global biodiversity models. An analysis of homologous primer site sequences obtained from whole genome data of prokaryotes and eukaryotes, revealing the presence of targeted Folmer primer sequences, was carried out. In addition, they demonstrated that several genomes of bacterial species had more closely paired nucleotides for the reverse primer than most animals.

In addition, DNA barcodes play a crucial role in the authentication of natural health products and the identification of medicinal plants. By verifying the botanical identity of herbal remedies and medicinal plants, consumers can be assured of product quality and efficacy, while helping to prevent mislabelling or adulteration of herbal products. Overall, the DNA barcode offers a versatile and indispensable tool for a wide range applications, contributing to diverse fields including biodiversity conservation, agriculture, public health, environmental monitoring and phytotherapy.

Species identification

DNA barcoding is a crucial tool for species identification, particularly where traditional methods are impractical. This is particularly useful in scenarios such as larval identification, where diagnostic characteristics are rare, and for linking different life stages of organisms. In addition, for CITES-listed species, barcoding techniques play an essential role in monitoring illegal trade, helping authorities to detect and prevent trafficking endangered species.

The detection of invasive species is another area where DNA barcodes are proving useful. At border control points, where rapid and accurate morphological identification may not be possible due to similarities between species or a lack of diagnostic characteristics, DNA barcoding offers a reliable alternative. This method enables authorities to rapidly detect invasive species in ecosystems and to differentiate between invasive and native species that may be morphologically similar, thus making a significant contribution to the management of biological invasions.

In addition, DNA barcoding facilitates the delimitation of cryptic species, which are genetically distinct but morphologically similar. By analysing DNA barcodes, researchers can uncover a hidden diversity species that might not be apparent from traditional morphological examination alone. However, the process of delineating cryptic species using DNA barcoding can be nuanced, as it depends on the analysis methods chosen. Despite the potential challenges, DNA barcoding provides valuable information on biodiversity and species boundaries, contributing to our understanding of the natural world.

Phytotherapy

DNA barcoding has become a powerful tool for identifying medicinal plants, offering a means of overcoming the challenges posed by adulteration and quality control in medicinal plant manufacture. In our recent review of 366 medicinal plant manuscripts published over the past decade, we found that DNA barcoding was widely used, with researchers utilising various barcode regions to study medicinal plant material. In particular, the Internal Transcribe Spacer 2 (ITS2) region has gained popularity due to its high discrimination rates at lower taxonomic levels. However, the adoption a multilocus DNA barcoding approach holds promise for more accurate identification of medicinal plants.

Despite the potential of DNA barcoding, there are limitations that need to be taken into account. The growing consumption of herbal medicines has led to an increase in adulteration during the manufacturing process, which has had an impact on the quality and safety of herbal products. Unethical practices such as misleading advertising, product substitution and contamination pose serious health risks to consumers. Regulatory agencies such as the FDA and CFIA face difficulties in adequately monitoring food supplement manufacturers and enforcing regulations designed to protect consumers.

Chemical research studies have highlighted problems associated with poor quality control and variability in the content of active ingredients among herbal supplements from different manufacturers. Studies based on DNA technology have identified contamination of herbal products by toxic plants, highlighting the need for robust authentication methods. DNA barcoding offers a promising approach to product authentication, providing a means of detecting substitution and contamination of plant-based products. Although it may involve some cost, bulk product testing using DNA barcoding can certify the authenticity and quality of herbal products, building consumer confidence and promoting transparency in the herbal industry.

Forensic medicine

DNA barcodes play a crucial role in forensic science, particularly in cases species identification. When unknown samples of animals or plants are discovered at crime scenes, a DNA barcode is used to identify them, potentially linking the evidence to a suspect and helping to secure convictions. This method is particularly relevant in crimes such as poaching, illegal killing of endangered species and animal abuse cases, where animal DNA is frequently used as evidence.

In cases of animal-related crime, DNA barcodes are a powerful tool enabling investigators to establish the identity of the species involved, thereby helping to prosecute offenders. Conversely, when plant DNA is detected as a trace, the DNA barcode helps link a suspect to a crime scene. By analysing the DNA profiles of plant samples found at the scene, forensic scientists can gather valuable information that can link individuals to criminal activity.

Overall, DNA barcoding enhances the capabilities of forensic science by providing reliable methods for identifying species from biological evidence collected at crime scenes. Whether animal or plant DNA, this technology is a valuable asset in forensic investigations, helping law enforcement agencies in their efforts to solve crimes and bring perpetrators to justice.

A DNA barcode database has been developed specifically for commercial timber tree species, serving as a reliable and secure tool to differentiate adulterant timber species for voluntary timber certification agencies. DNA barcoding methods have been used to identify timber species such as Santalum and Dalbergia, helping to prevent illegal trade in these woods.

Exposure to poisonous plants presents a significant health risk, requiring accurate species identification to determine the appropriate treatment. Visual observation of ingested plants is often impractical due to their degradation, making DNA barcoding an invaluable molecular tool for species identification. Studies have shown that the rbcL DNA marker can be used as a primary marker to identify venomous plants . In cases where higher resolution is required, ITS2 or trnH-psbA can be used as secondary markers. Some studies are laying the foundations for the development of a reliable molecular method for identifying toxic species vomit samples in cases poisoning, thereby contributing to rapid and accurate medical intervention.

Food industry

Food additives derived from plants or animals are widely used in the food industry as preservatives, antimicrobials, nutrition enhancers and colours to improve sensory attributes, food safety and quality. As consumer demand for healthy food increases, natural additives are frequently used not only as natural seasonings, but also for their medicinal properties, such as antihypertensive, antimicrobial, antioxidant, anti-inflammatory and anticancer effects. As a result, the use of medicinal plants as food additives has become increasingly popular around the world due to their health benefits. However, adulteration remains a major concern when it comes to food additives. Incorrect labelling or falsification of food products can lead to serious health problems, including cross-contamination and allergic reactions. More specifically, the falsification of meat products presents a major health risk of disease transmission. In addition, misidentification of food ingredients can cause allergic reactions in sensitive individuals. Food fraud has therefore become a critical risk factor for food safety. There is therefore a growing need for authentication and traceability to improve food safety. Mitochondrial DNA offers a promising avenue for species authentication in food and food products. Effective DNA barcode markers have been developed to accurately identify species, leading to approval by regulatory bodies such as the US FDA for use in a variety of food products. The use of DNA barcode technology as a regulatory tool enables precise identification and guarantees the authenticity of food products. By using DNA barcodes, the risks associated with microbiological and toxicological contamination of foodstuffs can be minimised. Ultimately, the DNA barcode has the potential to become the reference method for guaranteeing food authenticity and detecting fraud.

Traditional medicine

A variety of animals, including mammals, reptiles, fish and amphibians, have long been used as materials in traditional medicine (TM), particularly in Asian and African countries. For example, the horns, scales, muscles and gall bladders of wild mammals are important sources TM ingredients. However, it is difficult to distinguish authentic materials from adulterants, as in the case of Saiga tatarica horns, whose population has rapidly declined due to hunting. To solve this problem, DNA bar-coding techniques have been used, precise identification of species origins and differentiation between genuine and counterfeit products. Similar authentication have been applied to other MT materials derived from mammals, notably pangolin scales and deer musk.

Reptiles, such as snakes, geckos and turtles, also play an essential role in traditional medicine worldwide, offering a variety of therapeutic benefits. However, identifying medicinal reptiles with similar morphologies can be difficult. DNA barcoding provides a reliable strategy for distinguishing species, as demonstrated by snakes such as *Zaocys ends* and *Multilayer bungalow*, commonly used in TM. Using a panel of full and mini-codes, including COI, 12S rRNA, 16S rRNA and Cytb, the researchers have succeeded in identifying specific snake species and differentiating genuine products from adulterants.

Similarly, DNA barcodes have played a key role in the authentication of fish-based TM materials, such as shark fins and seahorses, which are often overfished and traded illegally. By using minibar codes, even from processed products, precise identification of species can be achieved, thus contributing to the conservation of endangered fish populations. Finally, amphibians, such as toads and frogs, are also commonly used in TM, presenting challenges species authentication. DNA barcoding techniques based on COI and 16S rRNA sequences have proved effective in distinguishing between different species of frogs and toads, thus guaranteeing the authenticity of amphibian-based TM materials.

DNA metabarcoding has become a promising tool for authenticating marked species and detecting undeclared taxa in TM formulations of animal origin. For example, studies using 16S rRNA barcoding sequencing and multilocus metabarcoding approaches have uncovered discrepancies between marked species and the actual content of various TM preparations. These results highlight the potential of DNA metabarcoding in pharmacovigilance for pre- and post-marketing audit of TM products. Interestingly, DNA metabarcoding has also found application in the identification of food composition from animal secretions, such as honey floral species, demonstrating its versatility in biodiversity analysis. Overall, DNA metabarcoding shows great promise as a robust tool for ensuring the safety, legality and authenticity of TM products, as well as for conducting biodiversity assessments in a variety of contexts.

CONCLUSION

In the biological sciences, DNA barcoding has become a transformative tool, revolutionising fields ranging from biodiversity conservation to forensic medicine. Through the analysis of standardised genetic markers, DNA barcoding enables rapid and accurate identification of species, addressing long-standing challenges in taxonomy, ecology and beyond. This comprehensive conclusion summarises myriad advances and applications of DNA barcoding, highlighting its importance, limitations and future prospects.

The evolution of DNA barcoding technology has been marked by significant advances, driven by innovations in molecular biology, bioinformatics and sequencing methodologies. Initial efforts focused on identifying universal genetic markers, leading to the selection of regions such as the mitochondrial cytochrome c oxidase subunit I (COI) gene and the chloroplast ribulose-1,5-bisphosphate carboxylase/oxygenase (rbcL) gene. Subsequent developments have extended the repertoire of barcode regions to include nuclear markers and whole genome sequencing approaches, offering greater resolution and accuracy in species identification.

The advent of high-speed sequencing platforms has revolutionised DNA barcoding, enabling the simultaneous analysis of several samples and the generation of large quantities of sequence data. This evolution has facilitated large-scale biodiversity surveys, metagenomic analyses and forensic investigations, accelerating the pace of discovery and innovation in DNA barcoding research. In addition, advances in bioinformatics tools and computer algorithms have improved data processing, storage and analysis, enabling researchers to extract meaningful information complex genetic datasets with unprecedented efficiency and accuracy.

The versatility of DNA barcoding extends to a wide range of scientific disciplines, each benefiting from its unique capabilities and applications. In the field of biodiversity conservation, DNA barcoding is a powerful tool for species identification, facilitating the assessment of species richness, population dynamics and community structure. By cataloguing genetic diversity and detecting cryptic species, DNA barcodes contribute to the conservation of threatened species and the management of invasive species, thereby promoting the health and resilience of ecosystems.

In the field of agriculture and food safety, DNA barcodes play an essential role in traceability and authentication, enabling the precise identification of plant and animal species in agricultural products and food supply . By detecting food fraud, adulteration and mislabelling, DNA barcodes help to protect consumer health and promote transparency and integrity in the food industry. In addition, DNA barcodes facilitate the identification of agricultural pests and pathogens, informing pest management strategies and reducing crop losses and agricultural risks.

Forensic science is another area where DNA barcoding has made a significant contribution, particularly in criminal investigations and forensic cases. By analysing DNA evidence recovered from crime scenes, DNA barcodes enable the identification animal and plant species, linking suspects to criminal activities such as poaching, wildlife and environmental crime. DNA barcodes also trace analysis, making it easier to solve criminal cases and administer justice.

Despite its widespread applications and advances, DNA barcoding is not without its limitations and challenges. Issues such as incomplete reference databases, genetic variation within species and technical constraints are obstacles to the widespread adoption and implementation of DNA barcoding in some contexts. In addition, ethical considerations, privacy concerns and regulatory frameworks must be carefully considered to ensure the responsible and ethical use of DNA barcoding technology.

Looking ahead, the future of DNA barcoding is extremely bright, fuelled by ongoing innovations in sequencing technologies, bioinformatics tools and analytical methodologies. Ongoing efforts to expand reference databases, improve taxonomic coverage and standardise protocols will improve the reliability, accuracy and applicability of DNA barcoding across a range of scientific disciplines. In addition, interdisciplinary collaborations, data sharing initiatives and public engagement efforts will promote greater awareness, understanding and use of DNA barcoding in research, education and policy making.

In conclusion, DNA barcoding represents a transformative paradigm in the biological sciences, offering unprecedented insights into the diversity, distribution and dynamics of life on Earth. With its broad applications, technological advances and interdisciplinary implications, DNA barcoding is poised to revolutionise our understanding of biodiversity, inform conservation efforts and address pressing societal challenges in the years ahead. As we continue to unlock the potential of DNA barcoding, we embark on a journey of discovery and innovation, guided by the promise of a more sustainable, equitable and resilient future for generations to come.

REFERENCES

CBOL - Plant Working Group. 2009. A DNA barcode for land plants. Proceedings of the National Academy of Sciences USA 106: 12794-12797.

Chase, M.W., Soltisn, DE, Olmstead, RG, Morgan, D., Les, DH, Mishler, BD, Duvall, MR, Price, RA, Hills, H.G., Qiu, YL, Kron, KA, Rettig, J.H., Conti , E., Palmer, J.D., Manhart, J.R., Sytsma, K.J., Michaels, H.J., Kress, W.J., Karol, KG, Clark, WD, Hedron, M., Gaut, BS, Jansen, RK, Kim, K.J., Wimpee , CE, Smith, JE, Furnier, GR, Strauss, SH, Xiang, QY, Plunkett, GM, Soltis, RS.., Swensen, SM, Williams, SE, Gadek, RA, Quinn, CJ, Eguiarte, L.E., Golenberg, E., Learn, G.H. Jr, Graham, S.W., Barrett, S.C.H., Dayanandan, S. and Albert, VA. 1993. Phylogenetics of seed plants: a nucleotide sequence analysis of the *rbcL* plastid gene. Annals of the Missouri Botanical Garden 80: 528-580.

Hebert, P. D. N., Cywinska, A., Ball, S. L. and deWaard, J. R. 2003. Biological identification using DNA barcoding. Proceedings of the Royal Society B: Biological Sciences 270: 313-321.

Hebert, P.D.N., Cywinska, A., Ball, S.L. and deWaard, J.R. 2003. Biological identification using DNA barcoding. Proc R Soc Lond B Biol Sci 270: 313-321.

Kurian A. 2019. Molecular systematics of south Indian rattans. These de
D. thesis, Cochin University of Science and Technology, Kerala, India.

Kurian A., S.A. Dev, V.B. Sreekumar, E. M. Muralidharan (2020) The low copy number nuclear region, RPB2 as a novel DNA barcoding region for species identification of the rattan genus *Calamus*
(Arecacees). Plant Physiology and Molecular Biology, 26: 18751887.

Mayr, E. 1969. Principles of Systematic Zoology. McGraw-Hill, New York, NY. 428 pp.

Newmaster, S.G., Fazekas, A. and Ragupathy, S. 2006. DNA barcoding in terrestrial plants: evaluation of *RBCL* in a multi-gene, multi-level approach. Canadian Journal of Botany 84: 335-341.

Thomas, M.M., Garwood, N.C., Baker, W.J.. 2006. Molecular phylogeny of the palm genus *Chamedoree*, based on the low copy number nuclear genes *PRK* and

RPB2. Molecular Phylogenetics and Evolution 38(2):398-415.

Vijayan, K. and Tsou, C. 2010. DNA barcoding in plants: taxonomy in a new perspective. Current Science (Bangalore) 99*:* 1530-1541.

Wilson, MA, Gaut, B. and Clegg, MT. 1990. Chloroplast DNA evolves slowly in the palm family (Arecaceae). Molecular Biology and Evolution 7: 303-314.

Wu, L., Wu, M., Cui, N.2021. Plant super-barcode: a case study on genome-based identification of closely related *Fritillary* species. *Menton Med* 16, 52

MIX
Papier aus verantwortungsvollen Quellen
Paper from responsible sources
FSC® C105338

FSC
www.fsc.org

Printed by Books on Demand GmbH, Norderstedt / Germany